YOUR KNOWLEDGE HAS VALUE

Phytochemical Screening of Stephania Abyssinica and Brassica Oleracea Var. Capitata L. Leaves and Evaluation of Antibacterial Activity

Kitesa Ayana

Bibliographic information published by the German National Library:

The German National Library lists this publication in the National Bibliography; detailed bibliographic data are available on the Internet at http://dnb.dnb.de.

ISBN: 9783346881199
This book is also available as an ebook.

Print and binding: Books on Demand GmbH, Norderstedt, Germany
Printed on acid-free paper from responsible sources.

The present work has been carefully prepared. Nevertheless, authors and publishers do not incur liability for the correctness of information, notes, links and advice as well as any printing errors.

GRIN web shop: https://www.grin.com/document/1360363

Phytochemical Screening Of Stephania Abyssinica And Brassica Oleracea Var. Capitata L. Leaves And Evaluation Of Antibacterial Activity

Kitesa Ayana

Department of Chemistry, College of Natural Sciences, Jimma University, Jimma, Ethiopia

ABSTRACT: Medicinal plants have played a crucial role as a source of traditional medicine in Ethiopia from time immemorial to combat different ailments and human sufferings. Leaf of Stephania abyssinica are used in Ethiopia as traditional medicine for the treatment of stomach aches, headaches, and inflammation. And Brassica oleracea L. var. Capitata (cabbage) has been widely used as an herbal medicine to treat gastrointestinal disorders and can also be used as food. This study aimed to carry out a phytochemical analysis and evaluation of the antibacterial activities of Stephania abyssinica and Brassica oleracea L. var. Capitata leaf. The leaves were collected and dried, powdered, and subjected to extraction using maceration technique. The results showed that there are alkaloids, flavonoids, glycosides, terpenoids, phenols, and saponins in the petroleum ether, acetone, and 50% chloroform in methanol leaves extracts of both plants. The leaf extracts of both plants were tested against E. coli, S. typhus, P.aerugenosa, and S. aureus. The results showed that the S. abyssinica crude extracts had better zone of growth inhibition against S. aureus and the acetone crude extracts of B. oleraceae var. capitatal had good activity against the four strains. Further isolation and characterization of pure compounds are recommended.

From the leaf extracts of S. abyssinica 2.7, 12.8, and 1.3 (% w/w) crude extracts were obtained from petroleum ether, and 50% chloroform in methanol and acetone, respectively. Similarly, 6, 2, and 0.4 (%w/w) crude extracts were obtained using petroleum ether, and 50% chloroform in methanol and acetone respectively were obtained from leaves of B. Oleracea Var. Capitata L. Hence, the results of the phytochemical analysis indicated that as there are alkaloids, flavonoids, glycosides, terpenoids, phenols, and saponins in the petroleum ether, acetone and 50% chloroform in methanol leaves extracts of S.abyssinica and B. oleraceae var. capital.The S. abyssinica crude extracts have shown better zone of growth inhibition against S. aureus 13±3.53, 12±3.46 and 11±3.39 mm and the acetone crude extracts of B. oleraceae var. capitatal have good activity against the four strains 12±1.65 and 13.5+3.39 mm against S. typhus and S. aureus. Its antifungal activity test and further isolation, and characterization of pure compounds are recommended.

Keywords: crude extracts, phytochemical analysis, Traditional medicine, Jimma

Contents

Introduction

Traditional medicine is the major source of treatment for large portions of human populations in developing countries. It is estimated that 80% of developing countries' population rely on traditional medicine, mostly plant drugs, for their primary health care needs[1]. Particularly in resource-poor communities, local therapy using traditional medicine is the only means of treatment.[2] Herbal remedies are becoming popular throughout the world because, though allopathic medicine can cure a wide range of diseases, its high prices and occasional side effects are causing many people to return to herbal medicines which tend to have fewer side effects.[3]

In Ethiopia, large populations treat different human and livestock ailments using traditional medicine using medicinal plants. Despite modern medicine becoming more widespread in the county, about 80 to 90 % of Ethiopia's population relies on traditional medicine to meet their primary healthcare needs.[4-6] The current healthcare system in Ethiopia is a primarily health care focused system that improves access to modern medicine more than ever.[7] However, the majority of populations continue to use traditional medicine. This is because traditional medicine is the most affordable and easily accessible source of treatment to the poor community[2] and cultural acceptance of traditional medicine.[8]

The use of medicinal plants as a source of traditional medicine has been inherited through generations in Ethiopia. It is an important component of the health care system in the county. The skills are however fragile and easily forgettable as most indigenous knowledge transfer in Ethiopia is based on oral transmission.9 With the current rate of modernization, it's logical to assume that traditional knowledge of medicinal plants is under threat of extinction.[1] Caring comprehensive studies are therefore important to document traditional knowledge on medicinal plant uses. Based on the above insight, the present study is aimed to evaluate the antibacterial activities of the two selected plants and underlines the importance of traditional knowledge used for the treatment of fire burn wound bacteria diseases in Gida Ayana woreda, Koneji kebele, East Wollega.

Materials and methods

Chemicals and Equipment

Chemicals used for extractions were petroleum ether, chloroform: Methanol and Acetone, and those used for phytochemical and antimicrobial activity tests were concentrated HCl, lead acetate solutions, water, concentrated H_2SO_4, ferric chloride solution, chloroform, Muller Hinton Agar (MHA), 4% Dimethyl sulphoxide (DMSO). The equipment used for extraction is a round bottom flask and rotary vapors that are used to concentrate the extracted at reduced temperature and pressure. Those that are used for phytochemical and antibacterial tests are beakers, pipets, measuring cylinders, test tubes, and discs.

Plant Collection

The dried leaf of *S. abyssinica* and *B. oleveraceavar.capitat L.* were collected from East Wollega, Gida Ayana woreda, and Jimma zone, local markets respectively. The air-dried leaves were collected. Then, the leaves were grounded into a fine powder using an electric grinder and stored at room temperature.

Extraction

The 63.3 g and 224.2g of air-dried powdered leaves of S.abyssinica and B. oleracea var. capitata L. were taken respectively and the samples were soaked in 500 mL of each solvent for 48 hr. Then the soaked samples were filtered out using Whatman No.1 filter paper. The filtrated was evaporated in a rotary evaporator at reduced temperature and pressure. The crude extract was collected in vials which were washed with detergent followed with acetone and stored at room temperature for further phytochemical test and antimicrobial activity.

Bacterial strains

For the antibacterial tests, four different clinical microbial isolates Gram negative and Gram positive were obtained and identified by using conventional biochemical tests and cultivated in pure culture, obtained from at Biology Department at the Microbiology laboratory, Jimma University.

Preliminary phytochemical screening

The methanol; chloroform, petroleum ether, and acetone extract of S. abyssinica and B. leverage var capitata l. leaves were screened according to Trease and Evans (1996), cited by (Musa et al., 2009), for its phytochemical constituents according to standard procedures.

Test for Terpenoids

Salkowski test: 5 mL of various solvent extract was mixed in 2 mL of chloroform followed by the careful addition of 3 mL concentrated H_2SO_4. A layer of the reddish brown coloration was formed at the interface thus indicating a positive result for the presence of terpenoids.

Test for flavonoids

NaOH test: 2 mL of the extract was dissolved in 10% aqueous NaOH solution and filtered to give yellow color, a change in color from yellow to colorless with the addition of dilute HCl indicated the presence of flavonoids.

Detection of Phenols

FeCl3 Test: Extracts were treated with 3-4 drops of FeCl3 solution. The formation of bluish-black color indicates the presence of phenols.

Test for Quinones

One milliliter of each of the various extracts was treated separately with FeCl3 solution. Quinines give coloration ranging from red to blue.

Test for Steroid

Salkowski test: A little quantity of the extract was dissolved in 1 mL chloroform and 1 mL of concentrated H2SO4 was added down the test tube to form two phases. The formation of red or yellow coloration was taken as an indication of the presence of sterols.

Detection of Saponins

Foam Test: 0.5 g of extract was shaken with 2 mL of water. If the foam produced persists for ten minutes it indicates the presence of saponins

Test for alkaloids

The extract (0.5 g) was stirred with 5 mL of 1% aqueous HCl in the water bath and filtered. Three ml of the filtrate was divided into three. To the first 1 mL, a few drops of Dragenroff's reagent were added and observed for the formation of orange to brownish precipitate.

Test for Glycosides

Formation of the brown ring at the interface by the addition of 2 mL of glacial CH3COOH followed by a few drops of FeCl3 solutions and 1mL of conc. H2SO4 to the extracts revealed the presence of glycosides.

Experimental procedure for in vitro-antibacterial activity test

Preparation of cultural media

Pure Muller Hinton Agar (MHA) was prepared by dissolving about 38 g of MHA in 1000 mL of distilled water and adjusted to pH 7.4 ± 0.2, sterilized by autoclaving at 121 0C for 15 min at 15 psi pressure, and used for sensitivity tests in the microbiology laboratory, Jimma University.

Preparation of Discs

From the plant extracts, 250 mg of each crude extract was dissolved in 1 mL of 4 % DMSO and 1 mL of the prepared extracts were taken and the filter Paper discs (Sterilized Whatman No. 1 filter paper discs of 6 mm diameter) were soaked into extracts to get 25 mg/disc Concentration and allowed to dry at room temperature.

Antibacterial Activity Test

Previously prepared paper discs containing different extracts were placed individually on the surface of the Petri plates, containing 25 mL of respective media seeded with 0.1 mL of previously prepared microbial suspensions individually (10 CFU/ mL). Standard antibiotic gentamycin (20 µg/ disc) obtained from Hi-media was used as the positive control. The discs containing n-hexane, dichloromethane, chloroform, and methanol solvents served as negative controls. The assessment of antimicrobial activity will be based on the measurement of inhibition zones formed around the discs. The plates were incubated for 24 hr. at 37°C and the diameter of the inhibition zones was recorded.

Result and Discussion

Percent Recovery of the Crude Extracts

From air-dried powdered leaves of S.abyssinica (63.3g) leaves, 2.7, 12.8, and 1.3% crude extracts were obtained from petroleum ether, and 50% chloroform in methanol and acetone solvents respectively. And from air-dried powdered B. oleracea Var. Capitata L (224.2 g) leaves1.6, 1.95, and 0.4% crude extract were obtained from petroleum ether, and 50% chloroform in methanol and acetone solvents respectively. The results indicated that S. abyssinica has a high amount of oil content than that B. Oleracea Var. Capitata L.

Phytochemical screening of the crude extracts

Table 1. Phytochemical screenings of crude extracts of *S.abyssinica* and *B. oleracea*

Phytochemicals	*S. Abyssinia*			*Brassica Oleracea*		
	P.ether	50%MeOH in Chloroform	Acetone	P.ether	50%MeOH in Chloroform	Acetone
Alkaloids	++	+	+	++	++	++
Flavonoids	+	+	-	+	+	+
Terpenoids	+	++	-	++	++	+
Saponins	-	++	-	-	-	-
Steroids	+	-	-	-	+	-
Phenols	-	++	++	-	-	-
Quinones	-	-	+	-	-	+
Glycosides	++	++	+	++	-	-

Key: ++ = high concentration + = moderate concentration, - = absence

As it has been tried to be indicated in Table 1 above, the phytochemical analysis of leaves extracts of B. Oleracea Var. Capitata L and S.abyyssinica have confirmed the presence and absence of different classes of secondary metabolites. From this, it shows the presence of alkaloids, and glycosides in all crude extracts of S. abyyssinica leaves, alkaloids also found in all extracts of B. Oleracea, and absence of quinones in both plant's crude extracts of petroleum ether, and 50% MeOH in Chloroform. The more intense color implies a higher concentration (++) and medium or normal coloration implies the presence of a sensible concentration (+). Therefore, phytochemical analysis helps to screen the phytochemical screening in a given medicinal plant. It has also taxonomical significance.

Antibacterial activity

Test Strains	S. abyssinica			Brassica Oleracea var capitata L.			control	
	Zone of Growth Inhibition in mm			Zone of Growth Inhibition in mm			positive	negative DMSO
	P.ether	50%MeOH in Chloroform	Acetone	P.ether	50%MeOH in Chloroform	Acetone		
E.coli	6.5±2.55mm	7±2.59 mm	6±2.5mm	6±2.61mm	9±2.88mm	8±2.79 mm	20mm	NI
S.typhus	11± 1.54mm	6.5±1.34mm	6.2±2.7mm	6.2±2.8m	11.5±3.2m	12±1.6 mm	25mm	NI
P.aerus	10±3.16mm	13±3.4mm	7±2.91mm	9±3.08 mm	11±3.24m	10±3.mm	29mm	NI
S. aureus	13±3.53mm	12±3.46mm	11±3.39 mm	6±2.8 mm	9±3.04mm	13.5±3.4mm	30mm	NI

Table 2. Antibacterial activities Test results of Leaves of S.abyssinica and B. Oleracea

Key: NI- No inhibition

6

The antibacterial activity of different solvent leaves extracts of Stephania abyssinica was summarized against four bacteria in Table 2. Petroleum ether extract of the plant showed a maximum zone of inhibition (13±3.53 mm) against S. aureus while a minimum zone of inhibition (6.5±2.55 mm) against E.coli. Chloroform: methanol extract of the plant also exhibited a maximum zone of inhibition (13±3.4 mm) against Pseudomonas aeruginosa while a minimum zone of inhibition (6.5±1.34 mm) against Salmonella typhus and acetone extracts shows a minimum zone of inhibition (6±2.5mm) against E.coli.

Conclusion

This research carried out phytochemical analysis and antibacterial activity evolution of S. abyssinica and B. Oleracea. The phytochemical analysis showed the presence of alkaloids, glycosides, saponins, and quinones in all crude extracts of both plants. The antibacterial activity test showed a maximum zone of inhibition against S. aureus and P. aeruginosa by 50%MeOH in Chloroform crude extract. The crude extract of both plants showed a high potential for antibacterial activities.

Recommendation

From the above conclusion the following recommendation was suggested;

✓ Isolation of the crude extract to find the active constituents that contributed to antibacterial activities

✓ Additional work is encouraged to elucidate the possible mechanism of action of these extracts and also the activity relationship of active constituents.

✓ Antifungal activity evaluation should be carried out

References

Mahmoud T, Gairola S. Traditional knowledge and use of medicinal plants in the Eastern Desert of Egypt: a case study from Wadi El-Gemal National Park. J. Med. 1 Plants Studies. 2013; 2 (6): 10–17.

Haile Y, Dilnesaw Y. Traditional medicinal plant knowledge and use by local healers in Sekrudistrikct, Jimma zone, Southwestern Ethiopia. J.Ethnobiol. Ethnomed. 2007; 3:24.

Kala CP. Current status of medicinal plants used by traditional Vaidyas in Uttaranchal state of India. Ethnobotany Res Appl. 2005; 3:267–278.

Baye H, Hymete A. Lead and cadmium accumulation in medicinal plants collected from environmentally different sites. Bull Environ ContamToxicol. 2010;84(2):197–201.

Bradley E, Thompson J, Byam P, et al. Access and quality of rural healthcare: Ethiopian Millennium Rural Initiative. Int J Qual Health C. 2011;23(3):222–230.

Elias A, Tesfaye G, Bizatu M. Aspects of common traditional medical practices applied for under-five children in Ethiopia, Oromia Region, Eastern-Harargie District, DadarWoreda. Journal of Community Medicine & Health Education. 2013; 3:6.

Samuel M, Leul L, Belaynew W, et al. Knowledge, attitude, and utilization of traditional medicine among the communities of Merawi town, Northwest Ethiopia: a cross-sectional study. Evidence-based complementary and alternative medicine. 2015.7.

Gakunga N, Mugisha K, Owiny D. et al. Effects of crude aqueous leaf extracts of Citropsisarticulata and Mystroxylonaethiopicum on sex hormone levels in male albino rate. International Journal of Pharmaceutical Science Invention. 2014;3:5–17.

Abebe D, Ayehu A. Medicinal plants and enigmatic health practices of Northern Ethiopia Addis Ababa: BerhaninaSelam Printing Enterprise. 1993.

Sendanayake L, Sylvester T, De Silva U, Dissanayake D, Daundasekera D and Sooriyapathirana S (2017) 'Consumer preference, antibacterial activity and genetic diversity of ginger (Zingiberofficinale Roscoe) cultivars grown in Sri Lanka' J. Agri. Sci., 12,. 3, 207-221.

LeenaTripathi, JaindranathTripathi. Trop. J. Pharm. Res. 2003; 2(2): 243-253.

T Syed Ismail, MSM Badhusha, M Mazhar. New Millennium seminar on medicinal plants proceedings 2002; 109-115.

8

MOA rekemase, RMO Kayode, AE Ajiboye. Int. J. Biol. 2011; 3: 3.

WHO. WHO Policy Perspectives on Medicines, Geneva 2002; 1-6.

AVWurochekker, EA Antony, W Obadiah. Afr. Biotech 2008; 7(16): 2777-2780.

Tropical   Plants   Database,   Ken   Fern.   tropical.theferns.info.   2019-05-30.
<tropical.theferns.info/viewtropical.php?id=Stephania+abyssinica>

Rokayya S, Li CJ, Zhao Y, Li Y, Sun CH. Cabbage (Brassica oleracea L. Var. capitata)
phytochemicals with antioxidant and anti-inflammatory potential. Asian Pac J Cancer
Prev. 2014; 14:6657–62.

Afolabi, F., Afolabi O.J, 2013. Phytochemical Constituents of Some Medicinal Plants in South
West, Nigeria. Journal of Applied Chemistry 4(1)76-78.

Rasigade JP, Vandenesch F. Staphylococcus aureus: a pathogen with still unresolved issues.
Infect. Genet. Evol. 2014 Jan; 21:510-4. [PubMed]

Chambers HF. Community-associated MRSA--resistance and virulence converge. N. Engl. J.
Med. 2005 Apr 07;352(14):1485-7. [PubMed]

Lancini, G., and Lorenzetti, R. 1993. Biosynthesis of Secondary Metabolites, in Biotechnology
of Antibiotics and Other Bioactive Microbial Metabolites. New York.

Nicolaou, K. C., Jason S. Chen, and Elias James Corey, 2011. Classics in Total Synthesis.
FurtherTargets,   Strategies, Methods III Weinheim: Wiley-VCH

Melton, L. 2006. "Body Blazes." Sci. Am. 294 (6): 24.

Appendix

Appendex 1: In vitro antibacterial activities of s.abyssinica and B.oleracea var capitata L leaves crude extracts

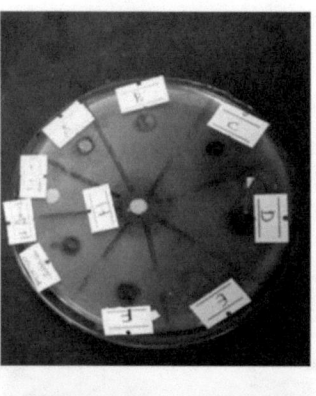

A= *E. coli*   B= *S.aureus*   C= *p.aeruginosa*

KEY

A= Acetone stephania abyssinica

B=P.ether Stephania abyssinica

C= Methanol: chloroform stephania abyssinica

D= Acetone B.oleveracea var capitata L.

E= P.ether B.oleveracea var capitata L

F= Methanol: Chloroform B.oleveracea var capitata L.

H= Gentamycin

Appendix 2: Phytochemical screenings of crude extracts of s.abyssinica and B.oleveracea var capitata L

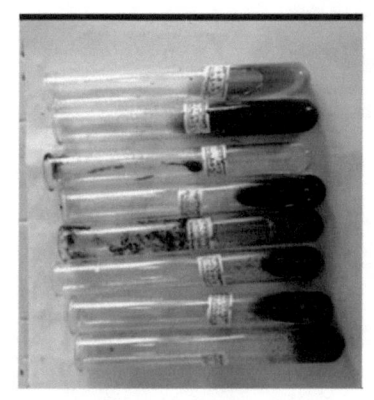
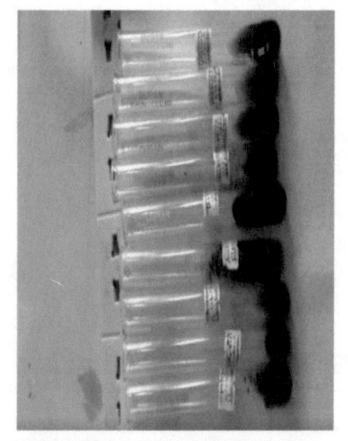

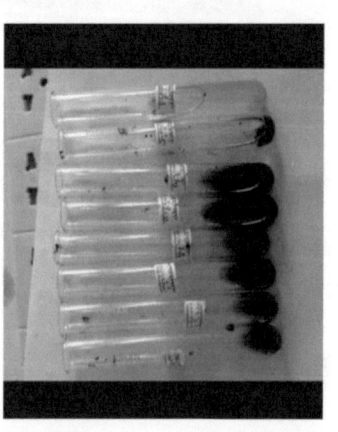

Appendix 3: Photograph of Brassica oleracea var. capitata L and Stephania abyssinica

Figure 2.1 Photograph of Brassica oleracea var. capitata L

Figure 2.2: Photograph of *Stephania abyssinica*

12